Cambridge **Primary**

Hodder Cambridge Primary

Science

Activity Book

C

Foundation Stage

Rosemary Feasey

HODDER
EDUCATION
AN HACHETTE UK COMPANY

The author and publishers would like to thank Chris Lawson, Science and Early Years Lead, Laurel Avenue Primary School, for her support in planning this material.

Orders: please contact Hely Hutchinson Centre, Milton Road, Didcot, Oxfordshire, OX11 7HH. Telephone: +44 (0)1235 827827. Email education@hachette.co.uk Lines are open from 9 a.m. to 5 p.m., Monday to Friday. You can also order through our website: www.hoddereducation.com

© Rosemary Feasey 2018

Published by Hodder Education

An Hachette UK Company

Carmelite House, 50 Victoria Embankment, London EC4Y 0DZ

Impression number   8

Year   2024

Cover by Steve Evans

Illustrations by Vian Oelofsen

Typeset in 17 pt FS Albert by Lizette Watkiss

Printed in the United Kingdom

A catalogue record for this title is available from the British Library

978 1 5104 4862 9

Contents

Naming and sorting food

 ✓ the foods you like. ✗ the foods you do not like.

bananas	egg	noodles	pizza
cheese	apple	doughnut	grapes
rice	tomato	bread	meat kebab

⭐ Circle the foods that are fruit.

strawberry

orange

grapes

banana

kiwi fruit

cheese

melon

egg

naan bread

spring rolls

coconut

Find out about food

We use our senses to find out about food.

| taste | hearing | sight | touch | smell |

 Choose some food to try. Draw and write about the food.

	The name of my food:
	It feels …
	It smells …
	It looks …
	It tastes …
	It sounds …

Tip:
Does it make a sound when you bite it?

Look inside fruits and vegetables

⭐ Join the fruits and vegetables to match a whole and a half.

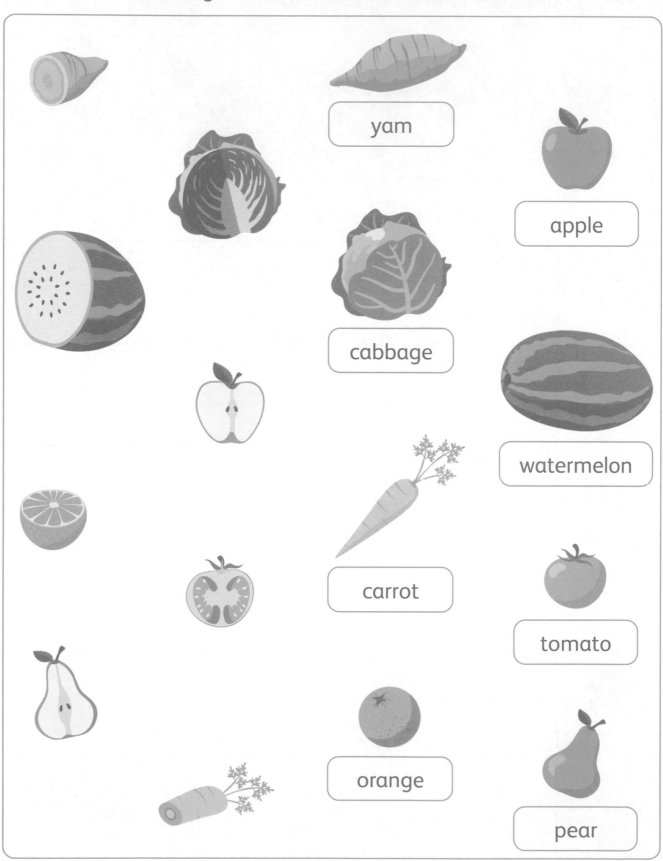

⭐ Choose a fruit or vegetable. Write its name.

I chose a _____ .

Use your senses to find out about the fruit or vegetable. How does it …

| look? | smell? | feel? | taste? | sound? |

⭐ Draw the outside of your fruit or vegetable.

The outside is _____ .

⭐ Draw the inside of your fruit or vegetable.

The inside is _____ .

More-healthy and less-healthy foods

 Min wants a snack. Help her to choose.

✓ more-healthy food.

✗ less-healthy food.

carrot sticks

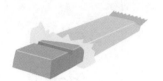

chocolate

orange

doughnut

ice cream

cake

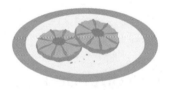

cookies

soda

banana

water

watermelon

cheese sandwich

8

These **raw** vegetables are a more-healthy snack!
Raw means not cooked.

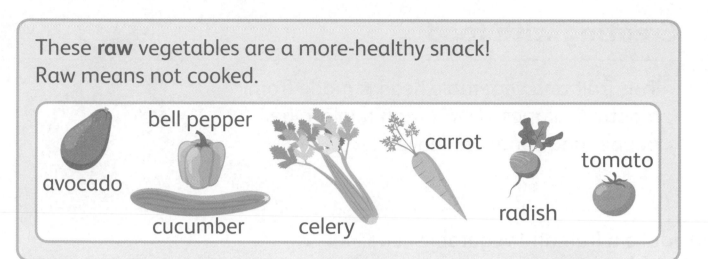

avocado · bell pepper · cucumber · celery · carrot · radish · tomato

⭐ Choose three raw vegetables to try.

Ask an adult to wash and cut up the vegetables for you first!

Draw and write the name of the vegetables.
Put a ✓ next to your favourite vegetable.

| I tried |
| _____. |
| I tried |
| _____. |
| I tried |
| _____. |

Creating with food

This fruit and vegetable head is made from a potato, grapes, a lemon, an apple, a bell pepper and coriander leaves!

⭐ Make a fruit and vegetable head.
Use toothpicks to attach the fruits and vegetables to the head.

I used these fruits and vegetables to make my fruit and vegetable head:

_____

_____

_____

My favourite vegetable is _____ because

_____.

⭐ Draw or stick a photograph of your fruit and vegetable head here.

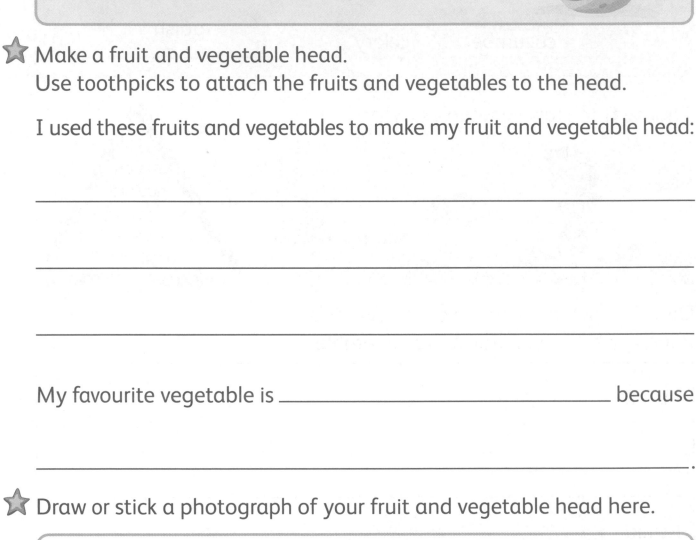

Design a new dessert. What will you use? Draw and label your dessert.

Ask an adult to wash and cut up the fruit for you first!

 Make your dessert.

 Taste your dessert! Write a sentence to say what it tasted like.

My dessert tasted _____.

Where does food come from?

Some food comes from animals.   Some food comes from plants.

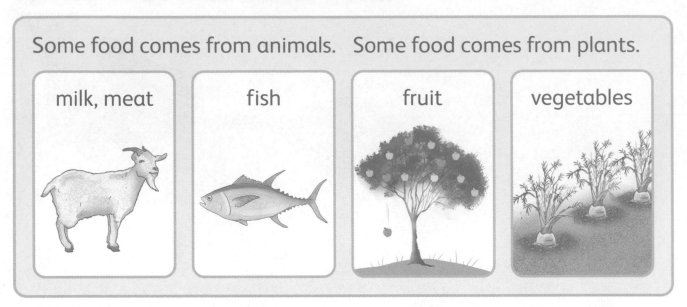

| milk, meat | fish | fruit | vegetables |

⭐ Draw a line from the food to where it comes from.

banana

potatoes

meat kebabs

tomatoes

bread

fish soup

eggs

rice

Draw your favourite meal on the plate. Label the food.

 Write where the food comes from.

The food in my meal comes from _____

_____

_____.

Growing food

Sabrina grew some mustard and cress seeds.

Sabrina ate the mustard and cress in a sandwich!

⭐ Choose some seeds to grow for food. Draw a picture to show how you planted the seeds. Add labels.

I planted _____ seeds.

How many days did your seeds take to grow?

_____

 Draw a picture of your plant.
Label the parts of your plant with these words.

| stem | leaf | roots |

 Try the food you have grown! Do you like it?

Ask an adult to wash and cut it up first!

Picnic time!

 Plan some food for a picnic. Draw and label the food on the plates.

 Put a ✓ next to the more-healthy foods. Circle the less-healthy foods.

⭐ Plan a test to find out the best way to keep a sandwich fresh.

Test each of these materials.

plastic box   paper bag   metal foil

Draw and label a picture to show how you will do it.

⭐ Write a sentence to show what you found out.

The _____

was the best at keeping the sandwich fresh because

_____ .

Getting to know dinosaurs

Dinosaurs do not live on the Earth today. They were once alive, but now they are **extinct**. This means that they have all died out.

⭐ Circle the animals that are extinct. ✓ the animals that are alive today.

You have a skeleton, and it is made of bones. Dinosaurs also had a skeleton made of bones. Some of these bones have been found, so we know what dinosaurs looked like.

 Join the labels to the tyrannosaurus skeleton.

back

tail

arms

feet

teeth

 Join the skeletons to the dinosaurs.

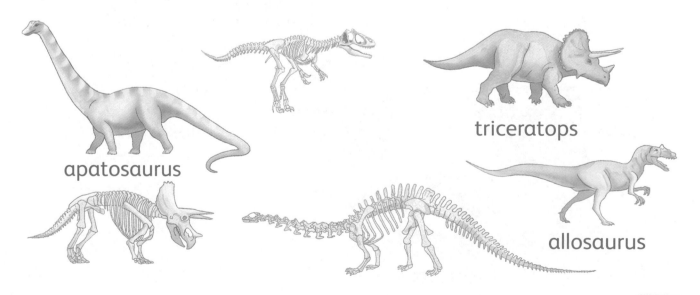

apatosaurus

triceratops

allosaurus

Comparing dinosaurs

⭐ Here is a triceratops. Join the body parts to the dinosaur.

| foot | eye | horn | mouth | tail | leg |

⭐ These dinosaurs both have a tail. Circle other things that are the same.

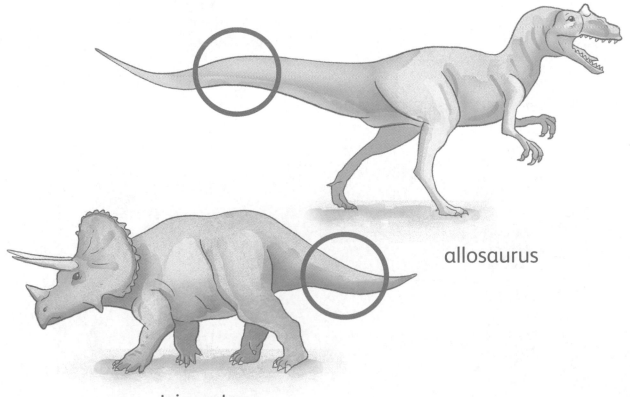

allosaurus

triceratops

The pterodactyl was a flying lizard, not a dinosaur. It lived at the same time as the dinosaurs.

 Put a ✓ to show what each animal has.

	2 legs	4 legs	wings	spikes
tyrannosaurus (ate meat)				
stegosaurus (ate plants)				
apatosaurus (ate plants)				
pterodactyl (ate fish)				
oviraptor (ate meat and plants)				

 Which is the odd one out? Say why.

Dinosaur habitats

The place where a dinosaur lived is called its **habitat**. Some dinosaurs lived in swamps, which were wet, muddy places. There was water to drink. There were lots of trees and plants to eat.

⭐ Choose a dinosaur. Find out about its habitat. Draw and label your dinosaur in its habitat.

Make a dinosaur swamp outdoors.

 Draw a picture or take a photo of your dinosaur swamp with toy dinosaurs in it.

I used these things to make my dinosaur swamp:

_____

_____

What did dinosaurs eat?

Draw three dinosaurs in this habitat to match the labels below.
Join each dinosaur to a label.

ate only plants

ate only meat

ate plants and meat

Dinosaurs had different teeth. Some dinosaurs had sharp teeth to eat other dinosaurs.

Some dinosaurs had flat teeth because they only ate plants.

 Here are some dinosaur skulls. Write about them.

I have _____ teeth.

I have _____ teeth.

I eat _____.

I eat _____.

I am a _____.

I am an _____.

brachiosaurus

allosaurus

flat   plants

meat   sharp

How big were the dinosaurs?

One of the tallest dinosaurs was a brachiosaurus. It was taller than a tree!

The smallest dinosaur was a lesothosaurus. It was the same size as a chicken.

 Was the dinosaur taller or shorter than a human?

Write [taller] or [shorter].

velociraptor	A velociraptor was _____ than a human.
allosaurus	An allosaurus was _____ than a human.
troodon	A troodon was _____ than a human.

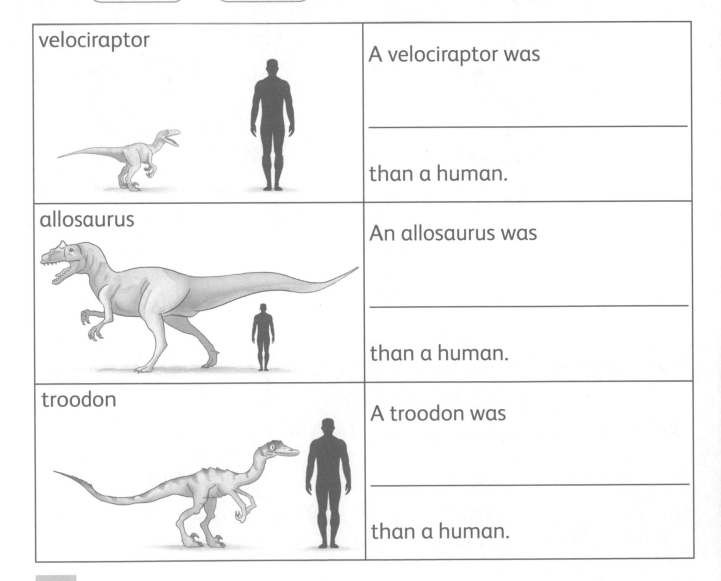

Fossils

A fossil is found in rocks. It is the parts of a plant or an animal that lived millions of years ago. Fossils help to show what dinosaurs looked like.

dinosaur fossil

⭐ What kind of fossil is it? Join the names to the fossil pictures.

| shell | bone | leaf |

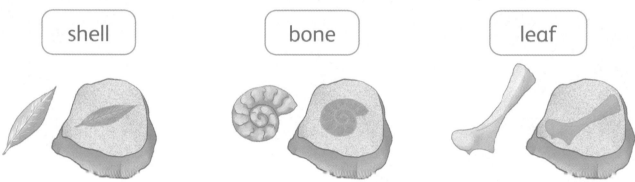

⭐ Join each dinosaur to the fossil of its footprint.

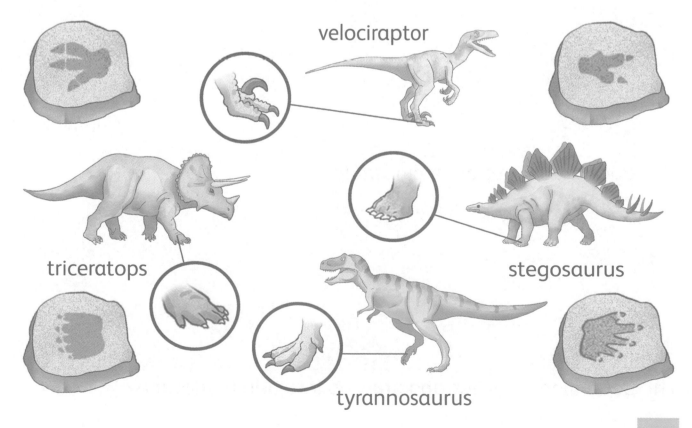

velociraptor

triceratops

stegosaurus

tyrannosaurus

Dinosaur eggs

Some dinosaurs laid eggs.
Their babies hatched from
the eggs.

 Draw a dinosaur life cycle.

The adult dinosaur lays eggs.

The eggs hatch into baby dinosaurs.

The baby dinosaurs eat and grow up into adult dinosaurs.

Design a dinosaur!

⭐ Design and make a dinosaur from junk materials.

⭐ Draw or stick a photograph of your dinosaur here.
Label the parts of your dinosaur's body.

The name of my dinosaur is _____.

⭐ Draw and write about your dinosaur.

My dinosaur's footprint

My dinosaur's nest and eggs

My dinosaur eats _____.

My dinosaur is bigger than a _____.

What can you remember?

 Join each sense to a picture.

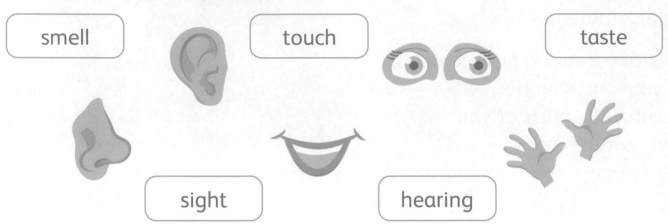

smell   touch   taste

sight   hearing

 Draw some food in each box.

more healthy ✓   less healthy ✗

 Circle the things that the mustard and cress seeds need to grow.

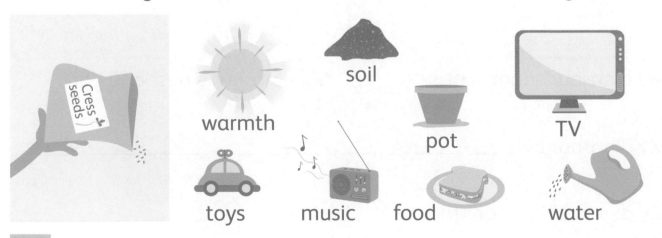

Cress seeds

warmth   soil   TV

toys   music   food   water

pot

⭐ Join the dinosaur to its name.

triceratops

allosaurus

tyrannosaurus

apatosaurus

⭐ Match each dinosaur to what it ate.

ate meat

ate plants

ate plants and meat

oviraptor

stegosaurus

tyrannosaurus

⭐ Circle the skull of the meat-eating dinosaur. Say why.

Self-assessment

Colour the stars to show what you can do!

Food	I can sort food into different groups.	☆
	I can say the names of different foods.	☆
	I can say what fruits and vegetables look like inside.	☆
	I can say how I use my senses to find out about different foods.	☆
	I can sort foods into more healthy and less healthy.	☆
	I can name some foods that come from plants and from animals.	☆
	I can say how to grow seeds.	☆
	I can do a test to find out the best way to keep a sandwich fresh.	☆
Dinosaurs	I can sort dinosaurs into different groups.	☆
	I know that dinosaurs were once alive, but are not alive today.	☆
	I can say how dinosaurs are different and how some are the same.	☆
	I know what extinct means.	☆
	I can name different dinosaurs.	☆
	I can name some dinosaurs that ate meat and some that ate plants.	☆
	I can say what a fossil is.	☆
	I can describe the life cycle of a dinosaur.	☆